CYBERSECURITY CHECKLIST

FOR BUSINESS OWNERS

Executive Battle Plan
to Survive Cyber Threats

By William Alexander Clements
CISSP, CCSP, CISM

COPYRIGHT

William Clements, wc@wclements.com, +1 (312) 313-0092

Published by divum llc. Manufactured in the United States of America.

ISBN 978-0-9762248-0-8 (Hardcover)
ISBN 978-0-9762248-1-5 (Paperback)
ISBN 978-0-9762248-2-2 (eBook)
ISBN 978-0-9762248-3-9 (Audiobook)

Cover photo courtesy of James Brodie.

LEGAL NOTICES

While all attempts have been made to verify information provided in this publication, neither the author nor the publisher assumes any responsibility for errors, omissions or contrary interpretation of the subject matter.

This publication is not intended for use as a source of legal or accounting advice. The publisher wants to stress that the information contained herein may be subject to varying federal, state, and/or local laws or regulations. All users are advised to retain competent counsel to determine what state and/or local laws or regulations may apply to the user's particular business.

The purchaser or reader of this publication assumes responsibility for the use of these materials and information. Adherence to all applicable federal/state/local laws and regulations, governing professional licensing, business practices, advertising and all other aspects of doing business in the United States or any other jurisdiction is the sole responsibility of the purchaser or reader. The author and publisher assume no responsibility or liability whatsoever on the behalf of any purchaser or reader of these materials.

Any perceived slights of specific people or organizations is unintentional.

To entrepreneurs and business owners that drive the world forward.

WILLIAM CLEMENTS

CONTENTS

WILLIAM CLEMENTS

INTRODUCTION

CYBERSECURITY RISKS CAN BE SIGNIFICANTLY REDUCED

As a successful business owner or manager, you have too much on your plate. The significant shift in cyber risks brings an avalanche of new challenges. You may not have the bandwidth for them. Luckily, you've picked up the right book.

This book is an action plan. In this book, we'll first get a baseline so we're on the same page. Then we'll immediately jump into action steps to improve your cybersecurity.

This book exists so you can:

- Understand the cyber risks your organization faces
- Determine the importance and urgency to assign to these risks
- Mitigate or transfer the significant risks

I want to help you protect your organization against cyber risk. It may seem daunting, but if we take a careful approach, we will get through it together.

First, you will identify the attack surfaces your organization has. Think of your business as a house. Attack surfaces are all the ways a burglar could try to break in—doors, windows, garage, or an unlocked side gate. In cybersecurity, an attack surface includes all the places where cybercriminals could try to access your systems or data. This could be anything from your employees' passwords to your company's website, email accounts, or even connected devices, like printers. The smaller your attack surface, the harder it is for hackers to find a way in.

Second, you will have an action plan to add defensive layers for these attack surfaces.

Third, you will have a checklist of steps you can take immediately, so you can jump-start your cybersecurity program.

- You can prevent most cyberattacks with solid security controls.

- You cannot prevent all cyberattacks. Build an incident plan and team in advance.

- Your cyber insurance carrier will fight your claim. Preparing evidence of proper business practices years before an attack is mandatory.

The owners and stakeholders at our client organizations can sleep well at night. They aren't worried about foreign adversaries hacking into their computer networks. They aren't stressing out about $250,000 wire transfers going to hackers. They are free to run and grow their businesses as if cybercrime isn't happening all around them.

I want the same for you: a clear head focused on your business objectives.

You got this. Your business will be secure.

Cybersecurity threats evolve every single day, and you need all the support you can get. To accelerate your progress and get exclusive resources that I could not put in this book, please visit:

https://cybersecuritychecklist.com/extra

Thank you for picking up this book. Now let's work on keeping your business out of the headlines.

Cybercrime is predicted to cost the world $10.5 billion in 2025.

(Cybersecurity Ventures 2024)

CHAPTER ONE

CYBERCRIMINALS NEVER SLEEP—BUT YOU CAN

Your organization is being targeted by cybercriminals. Every day. The only question is, how much damage will they be able to do?

In this book, I will include some stories from actual clients instead of only speaking in hypotheticals. In fact, let's talk about a client now. We have a client in the food supply chain who did not invest in any cybersecurity controls and knew they were at risk. This space is a prime target for international crime organizations. Criminals know companies will pay large sums to get back online if an attack causes a disruption. This business handles food worth hundreds of millions of dollars. It has a short shelf life. Every minute matters.

We started with an analysis and business risk review, so we could clearly outline the known risks. This includes both internal and external penetration tests. Think of penetration tests, or "pen tests," as hiring a professional to try and break into your business to find weak spots before the bad guys do.

- External Penetration Test: This simulates an attack from someone outside your business, like a hacker trying to break into your systems through the internet. It focuses on things like your website, public-facing systems, and internet connections.

- Internal Penetration Test: This assumes the attacker is already inside your network, such as a criminal who has already successfully compromised one computer inside your network. It checks how easily they could move laterally around your systems.

Both types of pen tests help spot vulnerabilities, so they can be fixed before any bad actors can take advantage of them.

As you can imagine, for this client, we advised implementing a full range of security controls.

Today, the stakeholders feel relieved because:

- We trained their teams to be vigilant.

- Their networks are locked down.

- Critical infrastructure is never reliant on a single device (we refer to this as redundancy).

- Their servers are secure.

- Their workstations only run known-good applications.

- Cybersecurity analysts are guarding their cloud and internal systems 24/7/365.

- Their financial controls will prevent wire transfers to criminals.

- If a criminal breaches their defenses, they can file a claim with their cyber insurer.

- We continue to conduct regular third-party audits to verify and document their cybersecurity.

We have another client in asset management. They had not been looking at cybersecurity as a business risk concern, so their historic security investment was low. The daily wire transfers of large sums made them prime targets for cybercriminals. Intercepting just one wire transfer would be a windfall for criminal organizations.

Cybercrime is expected to be more profitable than the global trade of all major illegal drugs combined by 2025.

(Bošković 2023)

During the initial security gap analysis, we uncovered over 30 significant risks. We advised that strong financial controls would be the best, easiest first step. For reference, here are some typical financial controls every business should implement:

- Dual Authorization: Require two people to approve all significant transactions, like large wire transfers or payroll changes. This reduces the risk of fraud, even if one account is compromised.

- Bank Alerts: Set up alerts with your bank for all transactions above a certain amount or for any suspicious activity.

- Separation of Duties: Divide responsibilities among employees so no single person has full control over financial processes, like issuing payments or reconciling accounts.

- Strong Vendor Verification: Always verify changes to vendor payment details with a phone call to a trusted contact. Email requests can be faked.

- Secure Online Banking: Limit access to your business's online banking to only the employees who need it, and always use multi-factor authentication (MFA) for all accounts (even accounts that only allow viewing transactions).

- Payment Limits: Set daily transaction limits with your bank to restrict the amount that can be withdrawn or transferred at one time.

- Cyber & Cybercrime Insurance: Invest in a policy that includes coverage for financial losses due to cyberattacks or fraud. Not all policies provide this type of coverage. Some policies provide very limited coverage in the event of an incident.

We also worked with their management to draft and enact a slate of company policies.

These measures, plus many technical controls, continue to assure stakeholders. No cybercriminal will have a major impact on this client's business.

We have another client in food processing. Before we worked with them, they had a decent cybersecurity budget but had very limited results. They had an executive mandate to maintain a strong cybersecurity program. They needed to carry cyber insurance since their largest customers require it.

Amazon sees over 750 million cyber attack attempts per day.

(Rundle 2024b)

We began with a comprehensive set of penetration tests and a business risk analysis. We also reviewed how they were using their current budget.

We soon discovered that they were not compliant with their cyber insurance requirements. This means that if an incident had occurred, they might not have had coverage.

We built a new cybersecurity program. It met their insurance requirements and created a paper trail to prove compliance. We also replaced their ineffective technical controls with a set of strong security solutions.

All owners and stakeholders are now confident. They believe that no cybercrime will harm their operations.

As you can see, a good cybersecurity program begins with an understanding of what you need to protect. Once we know that, we can build out a few layers of protection for everything we can. If your defenses are breached, we want to know as soon as possible. For that, we implement monitoring. As a last defense, you should invest in cyber and cybercrime insurance to help with the financial impact of a breach. To ensure your coverage, we document your compliance with cyber insurance requirements.

INCIDENT REPORT

On September 12, 2023, MGM Resorts International, a well-known hotel and casino company based in Las Vegas, became the victim of a major cyberattack. The attack compromised the computer systems of several MGM properties, including iconic casinos like the MGM Grand, Mandalay Bay, and Bellagio.

The hackers were able to gain access to sensitive customer information, including names, addresses, phone numbers, email addresses, credit card details, and even bank account information. This breach exposed the personal data of countless individuals, putting them at risk for identity theft and financial fraud.

According to MGM, the attackers exploited a security vulnerability in the technical support workflows to gain access to a computer system.

Upon discovering the breach, MGM Resorts acted swiftly to secure its systems and mitigate further damage. Despite these efforts, the incident took a significant toll. The total cost of the cyberattack is estimated at $110 million, encompassing direct costs like investigation, system repairs, and compensation. This amount does not include reputational impact (Rundle 2024a).

While this attack targeted one of the largest gaming corporations, the same type of attack was successful against a small staffing agency. Despite being a small business, they had the personal details of thousands of current and past temporary employees. Due to the ransom expense, alerting employees of the personal data they exposed, and settling claims with their current and past employees, the attack nearly destroyed them overnight.

CHAPTER TWO

CYBERCRIMINALS ARE TARGETING YOUR IT TEAM AND YOUR BUSINESS

As a Managed Service Provider (MSP), we have managed IT systems for businesses since 2000. (I founded the company when I was sixteen years old.) In the early days, although we had security risks, mitigation was straightforward. We used long, complex passwords. We only provided team members with the access they required and nothing more. We had a handful of best practices and followed them closely.

Today, each client has extremely confidential information, like administrative passwords and vital documents. As the IT firm managing these clients, we store all their confidential information. A breach of our systems could have a ripple effect. It would allow breaches of some clients' systems.

You can imagine how dangerous a breach of our systems would be.

You can imagine the consequences for our clients and our business as a result.

IT firms like ours have always been targets, but the bull's-eye is larger than ever before.

We've grown and taken on more clients. Our clients have grown. We're now managing more locations, networks, and computers. The volume of confidential information we're managing is higher than ever before. The size of the team that needs access to this confidential information is ever-increasing. The number of vendors we must share some of this information with is also at an all-time high.

Ensuring the right team members have exactly the right access is complex. Keeping access away from both those who don't need it and criminals is even more complex. Then there are the magnitudes of risks posed by departing team members.

Over the years, cybercrime has grown exponentially. It funds the world's largest criminal organizations. The risks from both inside and outside our company continued to grow.

The average cost of a data breach is $4.88 million, the highest total ever.

(IBM 2024b)

Now you see that our business faces significant cybersecurity risks, likely more risk than your business. The consequences of just one breach could be dire.

But our own security processes are rigorous:

- We have a third-party firm assess our security and controls every 90 days.

- We check security gaps and add new defenses. These include both administrative and technical controls.

- We review our internal training and add more as needed.

- We ensure our cyber insurance policy covers all known threats and consequences. We also document evidence of compliance with all policy requirements.

We must take our own cybersecurity seriously. We must constantly improve.

Unfortunately, many MSPs have experienced significant breaches. In one case, an MSP was compromised twice, and each time, most of their larger clients were impacted as well. It's the nightmare scenario every business fears—and it's enough to make anyone question their reliance on external providers. While I knock on wood, let me reassure you that we never had a breach.

INCIDENT REPORT

Lurie Children's Hospital of Chicago fell victim to a sophisticated cybersecurity attack in January 2024. Many major systems were offline for months, including their email, phones, health records, and patient portal.

The breach affected an estimated 800,000 people, with Personally Identifiable Information (PII) compromised including names, addresses, dates of birth, Social Security Numbers, and medical records. One criminal ransomware group claimed responsibility and allegedly sold the data for $3.4 million.

Lurie had to establish an entire call center and work with the FBI in the wake of the attack. Lawsuits are pending (Camarillo 2024, Washburn 2024).

It often takes a team of technically competent criminals to devise an attack against an organization like Lurie. Individual criminals, without any brilliant plan, successfully attack small dental and doctor practices every week. It is somewhat helpful that the news media reports on major cyberattacks because they include details about methods being employed, the extent of damage, and possible restitution for victims. But they can give the impression that these attacks only happen to large organizations. Most smaller businesses devastated by cyberattacks never make the news, even though they are most of the targets.

But fear doesn't have to rule your decisions. At my organization, we faced similar challenges until we built a strong cybersecurity infrastructure supported by a continuous review process. We know we've implemented the best policies, procedures, and services in the industry. While no system is completely immune, the risk of a breach is incredibly small. And if something does happen? We'll catch it right away.

For decades, I've been managing technology across businesses, refining processes to ensure they're as strong as possible. It's never a one-size-fits-all solution—it starts with a deep analysis of the business environment. We bring in penetration testers to identify security gaps, evaluate administrative and technical controls, and build a plan that closes those gaps. It's not an overnight transformation; it's a deliberate, step-by-step process.

At the end of that process, every stakeholder can confidently say, "We've done everything reasonable to protect ourselves." Any remaining risk is minimal—and more importantly, it is risk the organization can afford to accept without jeopardizing its future.

This is the place I want you to reach with the help of this book. I want you to sleep well at night, knowing you've taken every reasonable step. Any remaining risk will be nominal—something you can live with. Trust me, you can get there, and I'm excited to guide you along the way.

Technology can feel overwhelming—even for those of us who live and breathe it. That's why the IT industry is filled with specialists who focus deeply on narrow areas of expertise. It's simply too vast for any one person to master every aspect.

As a business owner or manager, you can't be an expert in both your industry and your business's cybersecurity. But even though you can't master it all, you need to have at least a basic understanding of the key risks you face. And just as importantly, you need a clear path to move your business from where it is today to a more secure place.

You might think you don't need to worry about this. I've got an IT person, either in-house or on contract, and they've got it all handled. But here's the truth: This is your business. The buck stops with you. You can outsource the tactical and technical work, but you can't outsource the responsibility.

Building your understanding won't take long—just a few minutes for a quick, solid check. Let's start here, and you'll see exactly where we're headed.

Please answer these questions as best you can. These are simple yes/no questions. No need to take any action at this point, even if an answer is no.

- Do you have a written inventory of your business's technology assets? These are all the systems and devices that could be compromised.

- Do you have a list of all the sensitive data your business stores? Knowing what data is at risk is critical.

- Does your team know how to recognize common cyber threats? Would they know how to spot phishing emails or suspicious activity and alert you right away?

- Is your leadership team aware of the company's overall cybersecurity readiness and the risks the organization faces?

- Does your IT team have a well-defined incident response process? This includes specific actions to mitigate threats, restore operations, and maintain service continuity.

- Do you have a communication plan in place for cybersecurity events? This ensures everyone knows the steps to take before, during, and after a breach—and how to stay updated throughout the process.

- Have you evaluated how a breach could impact your bottom line? It's essential to understand the potential financial consequences.

- Do you have a comprehensive insurance policy? It should cover not only incident-related expenses but also losses like the impact on customers and business operations.

- Are you confident that your business can survive a major cyberattack—even one with significant financial impact?

Even if some of these questions seem difficult to answer right now, we'll tackle them together and make sure you're ready for anything.

CHAPTER THREE

RETHINK YOUR SURVIVAL PROBABILITIES

Let's consider a handful of additional questions:

- Do you know which systems and software in your business could impact security?
- Are you comfortable delegating cybersecurity tasks, knowing the right processes are in place?
- How important is it to you that you can trust the technical controls protecting your business, so you can focus on what really matters?

Keep in mind that cybersecurity isn't like flipping a light switch—there's no instant fix. Risks evolve, and security is an ongoing process. The goal is to jump-start where you are and get your business moving toward a more secure state.

Another challenge you might face is not knowing which protection you already have in place. You may have systems, controls, and tools working in the background, but you're unsure how they all fit together. That's okay, too.

In fact, most businesses with fewer than 100 employees don't have in-house IT staff. Even large companies often outsource parts of their IT management and operations. Why? Because managing technology infrastructure and security is challenging. Cybersecurity requires experts, who are prone to burnout. High turnover in security roles is common, making it hard for many businesses to maintain the right talent in-house.

Let's also be realistic: Cybersecurity costs more than ever today. And much of that increase comes from the need to stay ahead of evolving threats. Instead of focusing on the cost of a cybersecurity program, though, I want you to consider the potential cost of not having one.

Imagine this: A cybercriminal gains access to your systems and transfers 99% of your operating account balance to a foreign bank, beyond the reach of U.S. authorities. How would that impact your business? Could you survive?

Or consider a ransomware attack: Criminals demand a ransom equal to your business's operating funds, knowing exactly how much you have. If you can't pay, they threaten to halt your operations indefinitely. What would you do? Would you have enough resources to continue operating without those funds?

One solution might be to set up a reserve account in a separate bank, with a different set of access controls and limited user permissions. Alternatively, you could secure a line of credit to keep operations running during a crisis. The key is to have a plan before disaster strikes.

Now, imagine another scenario: A cyberattack disrupts your business operations entirely. Your systems are down, and you can't process orders or serve customers. Everything grinds to a halt. How long could your business survive in that state?

If your operations aren't heavily reliant on technology, maybe you could move to manual processes for a while. In that case, you're in a stronger position to survive and recover from an attack. But for most businesses, shutting down their systems—even temporarily—isn't an option.

So, ask yourself: What's your limit? How many hours, days, or weeks could your business stay afloat without access to key systems? This is the kind of planning that separates businesses that survive cyber incidents from those that don't.

One final consideration is the reputational impact of a security breach. Imagine that government regulations require you to notify all your past and current customers that your business was compromised and their information ended up in the wrong hands. What would the impact on your business be?

For some businesses, the impact might be minimal—particularly if they are not heavily digitized. If your operations are mostly offline, your customers might shrug it off and move on. However, if you handle sensitive information—such as personal health data—the consequences could be far more severe. A breach of trust in a medical practice, for example, could lead patients to switch providers, potentially causing lasting financial and reputational damage.

When considering the potential risks and their impact, you need to weigh the cost of prevention. What would it cost to mitigate each of these scenarios? In some cases, it might not be worth the investment. If the impact of a breach is insignificant, spending a fortune to prevent it doesn't make sense.

Cyber insurance is one way to transfer some of this risk to a third party. However, it's important to understand that insurance carriers aren't going to take unreasonable risks. Many insurers now require businesses to have strong cybersecurity controls in place before offering a policy. If your cybersecurity posture isn't robust, insurers won't see you as a good bet—and they'll either deny coverage or charge steep premiums.

Think of it like homeowner's insurance. You can't expect an insurer to cover a home if you refuse to lock your doors and windows. Cyber insurance works the same way. If you take cybersecurity seriously, put strong controls in place, and follow best practices, you'll be in a good position to obtain a policy. But if you neglect your cybersecurity, no insurance provider will take that risk.

Here are some other questions to help you assess your current situation:

- Are you aware of your business's biggest cybersecurity risks?

- Have you implemented a slate of cybersecurity controls assembled by a cybersecurity consultant?

- Do you currently have a cyber insurance policy, or are you planning to apply for one?

- Have you gone through the process of answering security-related questions required for insurance approval?

If you don't have a cyber insurance policy, you need to take cybersecurity even more seriously. It is critical to close as many security gaps as possible to prevent a single attack from crippling your business.

But depending on your industry and the controls you already have in place, you might be closer to meeting insurance requirements than you realize. Later in this book, we'll discuss some essential cybersecurity controls to help you get on the right path.

Your goal should be to protect your business from ruin by a single cyberattack. If you do choose to apply for coverage, your preparation will put you in a strong position to secure a policy.

INCIDENT REPORT

Shan Hanes, former CEO of Heartland Tri-State Bank, allegedly wired $50 million to cyber criminals in a very successful pig butchering scam. This amount was significant enough to cause the bank to fail in August 2024. He also faced criminal charges for fraud, even though the criminals were likely the only beneficiaries of his actions (Office of the Inspector General, Board of Governors of the Federal Reserve System 2024).

Pig butchering is a cruel scam where criminals "fatten up" their victims before taking everything. It often starts with a friendly message—maybe someone pretending to be a long-lost friend, a romantic interest, or even a business opportunity. Over time, they build trust and slowly convince you to invest in what seems like a great deal, often using fake apps, websites, or cryptocurrency schemes.

Once they've gained your trust and you've invested a significant amount, they "butcher the pig" by disappearing with your money. It's called pig butchering because, like fattening a pig before slaughter, the scammers groom their victims before stealing as much as possible.

Many pig butchering targets are just average, hard-working folks who happen to have a 401k or some retirement savings. Anyone who uses the internet for communication can easily be targeted for this kind of crime. Mr. Hanes was an anomaly here and not a typical target.

CHAPTER FOUR

START WITH WHAT YOU KNOW

At this point, you might be wondering: What exactly needs to be done to secure my business? What services do I need? How do I know if I've covered all my bases?

Let's take a quick assessment before we move on:

- Do you have a list of where you believe your technology may be vulnerable?
- Do you know what you have in place or need to implement to protect these technological components?
- Do you know how you'll respond if a security incident occurs?
- Do you know what cybersecurity requirements your regulators, suppliers, customers, or industry groups have for your organization?

These questions form the meat and potatoes of cybersecurity for small businesses. Cyber threats are a reality for everyone, regardless of industry. While your business might not face the same risks as others, getting the fundamentals right is crucial.

And it all starts here. This is the core of your cybersecurity strategy, and from here, we'll build on that foundation together. With the right approach, you'll be well-prepared to protect your business from whatever challenges come your way.

Most clients have invested in antivirus software. They believe it is the cornerstone of cybersecurity programs. Many business owners thought antivirus alone was enough. That was the extent of their security efforts.

This leads us to the number one challenge businesses face in cybersecurity: they don't know what they don't know. This knowledge gap puts businesses at risk. They are vulnerable to unknown threats.

As you can imagine, building a comprehensive cybersecurity program takes time. On average, it takes nearly a year to build a strong security framework. It must address both technical and business risks. Cybersecurity requires continuous effort and investment, and leadership must be actively involved throughout.

CHAPTER FIVE

CYBERSECURITY CHECKLIST

The core of this framework is a checklist. You will check off the technology or attack surface you have in place, to the best of your knowledge. It is straightforward. Simply mark the ones you believe you have from what you know about your business and technology. For most, you will just check off the obvious technologies your organization is using.

For each technology or attack surface, you will have a general idea of the business risks and ways to reduce the risk. The business risk is provided so you understand the potential impact on your organization if the technology is compromised. The ways to reduce the risk are listed so you can competently work with your IT service providers and a cybersecurity professional to close any security gaps.

Like all other aspects of your business operations, this will always be a work in progress. Depending on your organization, you may want to go through this process monthly or quarterly. If any technology changes, you may want to go through it immediately after the change.

As a cybersecurity professional, I recommend seeking the guidance of an expert. They will be able to determine your security needs and build a cybersecurity program.

The challenge is, without clear details on each business risk, it can be hard to know whether a cybersecurity expert is providing exactly what you need.

This book exists for this very reason. I aim to equip every entrepreneur and leader with the knowledge to survive today's cyberwars. It is not enough just to engage a professional or to have an in-house team. You need to know the game plan. You need to know what is really at risk. You need to know your organization can survive and thrive in the face of these new threats.

Some elements of this list may require you to ask questions of others: team members, CPAs, IT, or vendors. But you can start with what you already know yourself. This will get the ball rolling. Then go through the list additional times and bring in answers from other people to expand.

Each step in the checklist is a technology or other potential attack surface. For each, you will have a ready-to-go plan of action to secure them.

As you use the checklist, follow the action plans for the attack surfaces you identified. You can disregard any attack surfaces you do not seem to have.

The implementation of administrative and technical controls requires expertise. You will want to delegate most of the steps listed. You will want to follow up and ensure that what they implement meets the guidelines listed.

To accelerate your progress and get exclusive resources that build on this checklist, please visit:

https://cybersecuritychecklist.com/extra

AUTHENTICATION & USER ACCESS RISKS

❑ My organization uses identity services, such as user accounts.

Business Risk

If your organization uses any system with user accounts, then you are using identity services. Most likely, you have several systems that provide these services, with one acting as the primary (such as a Windows Server, Microsoft 365, or Google Workspace). Identity services are critical because they authenticate users and establish what access each user has to different systems and resources.

Thirty years ago, securing your accounts was as simple as choosing a strong password. However, times have changed dramatically. Today, having a complex password alone is no longer sufficient. You need to account for the business risk associated with a criminal logging into a system using a stolen password.

> **In the Americas in 2023,**
> **51% of organizations first learned**
> **of a compromise from an external source,**
> **while 49% identified evidence**
> **of a compromise internally.**
>
> (Mandiant 2024)

The potential impact of unauthorized access varies by system. For example, if a criminal gains access to your Microsoft 365 environment, the risk is nearly limitless, as it could compromise your entire organization. In contrast, access to a basic expense tracking system might pose a lower risk. The key is to evaluate the potential damage if an attacker were to gain access.

Action Plan

- Establish strong password policies. Every system that provides identity services must enforce a strong password policy. This includes requiring complex passwords that are difficult to guess. Additionally, you should have a corporate policy that employees read and agree to, reminding them to choose secure passwords. Avoid common mistakes like using children's names, pets' names, or birthdays.

- Although difficult or impossible to enforce, employees should be reminded that each password needs to be completely unique and unrelated to other passwords. Password reuse is a major risk because once a criminal has an employee's password for one system, they will attempt to use those credentials on every major web-based application quickly.

- Implement MFA. All identity systems should require

MFA. While it might be a minor inconvenience for users, the security benefits far outweigh the hassle. Remember, even if a criminal gains access to a seemingly insignificant system, they could still gather valuable reconnaissance information. For instance, if an attacker gains access to a simple expense reporting system, they might be able to collect a list of current and former employees. This data could then be used to craft targeted phishing attacks over time. Let's not give attackers that opportunity.

- Enable monitoring and alerts for suspicious activity. Whenever possible, identity systems should have active monitoring and alerting to detect suspicious activity. Major platforms like Microsoft 365 and Google Workspace can be monitored by a Security Operations Center (SOC) for real-time responses to potential threats.

- For smaller systems, where SOC monitoring may not be an option, configure basic alerts that notify users when their accounts are accessed from new devices or different locations. Even a simple email alert can help users detect unauthorized access early.

- Require team members to use an enterprise-grade password manager. Storing passwords in Google Chrome, Microsoft Edge, or Mozilla Firefox is a dangerous practice. An enterprise password manager will keep credentials in a separate, secure location that requires a separate password. This will require some training and a strong push to convince everyone it is worth it, but it absolutely is worth the extra steps.

- Dark web monitoring is not a fast solution for preventing a breach, but it is helpful. This is a service where a company monitors usernames and passwords being sold or distributed on the dark web (by criminal groups). When they find credentials that belong to someone at your company, you will be alerted. Of course, if the criminal groups already have these credentials, damage may have already been done. It does help keep team members engaged because if they are notified when a password is exposed, it will encourage them to take password security more seriously moving forward.

- Implement Single Sign-On (SSO) for as many applications as possible. Especially those accessible from the internet from anywhere, you should implement SSO. There are two primary benefits for this: First, it makes it very easy for your IT team to disable a departing team member's access from a single, central interface. Second, if all or most of your applications use the same identity system, it means your security team is able to focus their attention on monitoring and responding to incidents from this single system.

- Many applications that do not provide direct security alerts may support SSO, so you can have the benefit of security monitoring with most mature web applications. Also, for applications that do not have built-in MFA functionality, using SSO with an identity system that has MFA means your applications also instantly have MFA. This also reduces the number of passwords and MFA mechanisms team members need to remember and maintain.

NETWORK INFRASTRUCTURE RISKS

❑ My organization uses a firewall at each physical location.

Business Risk

The firewall is akin to having someone at the reception desk. The purpose is to help direct traffic coming to and going from your network. While the firewall cannot prevent all risks from the outside world, it can provide some very fundamental security. This is an essential component. Without a business-grade firewall, your entire business is at eminent risk of attack from all attackers at any skill level.

Action Plan

- If you do not have business-grade firewalls, have a firewall installed as soon as possible (i.e. Meraki MX, SonicWall, or Fortinet FortiGate). If you have a firewall, but it is built into the modem/router from your Internet Service Provider, it is not adequate.

- Ensure you have a subscription for software updates for every firewall and that the software is updated.

- Ensure you have a person or team professionally managing these firewalls and updates.

❑ My organization uses network switches.

Business Risk

Network switches run software. All software can be compromised. Since all of your business information travels through these network switches, a compromised switch could theoretically provide an attacker with complete access to all confidential records.

Many passwords and other confidential records are sent back and forth across a network. Also, instructions for computing processes may be sent over the network.

A compromised switch could provide an attacker with access to nearly everything of value to your organization.

Action Plan

- Network switches should be managed and business grade. Unmanaged or consumer-grade switches should be replaced.

- Ensure you have a subscription for software updates for every network switch and that the software is updated on a regular basis.

- Ensure you have a person or team professionally managing these switches and updates.

- Depending on the complexity of your network, you may also benefit from having Virtual Local Area Networks (VLANs) and additional security services deployed. This would allow you to prevent communication between different segments of your network. For example, you could prevent a computer in a warehouse from talking with a computer in the finance office, since they likely do not need any direct communication. This reduces the risk of one compromised device being used as a point of access for other devices.

- Only purchase new equipment from trusted vendors and trusted manufacturers. You should also consider the country of origin.

INCIDENT REPORT

A Florida-based dual citizen of the United States and Türkiye named Onur Aksoy was recently sentenced to six and a half years in prison for orchestrating a massive operation to traffic counterfeit Cisco networking equipment. Over the course of several years, Aksoy's scheme funneled hundreds of millions of dollars' worth of low-quality and fraudulent devices into the U.S. supply chain, impacting critical sectors such as healthcare, education, and national defense.

As part of his plea agreement, Aksoy agreed to pay $100 million in restitution to Cisco, with additional compensation to other victims pending further court decisions. Authorities also seized and plan to destroy millions of dollars' worth of counterfeit goods linked to his operation.

The scale of Aksoy's activities was staggering. According to the U.S. Department of Justice, his network involved at least 19 shell companies based in Florida and New Jersey, as well as over two dozen online storefronts on platforms like Amazon and eBay. These entities imported tens of thousands of counterfeit devices from suppliers in China and Hong Kong, all falsely branded with Cisco trademarks to appear authentic. The devices, sold as high-quality networking equipment, had an estimated retail value in the hundreds of millions of dollars, generating over $100 million in revenue for Aksoy.

The counterfeit devices found their way into highly sensitive environments, including U.S. hospitals, schools, and even military systems supporting advanced fighter jets and aircraft. Such counterfeit products not only defrauded businesses but also posed significant risks to national security and public safety. Principal Deputy Assistant Attorney General Nicole M. Argentieri highlighted the dangers posed by counterfeit goods in critical systems, calling this case one of the largest counterfeit trademark prosecutions in U.S. history (U.S. Department of Justice 2024).

❑ My organization uses remote access.

If anyone at your organization or a third party can access any computers from outside your location, you have remote access technology in use. This could be with a Virtual Private Network (VPN) or another technology.

Business Risk

Any remote access technology poses a risk because it provides a potential way for an attacker to easily access devices on your internal network. For example, let's say a bookkeeper works from home some days and has remote access to the bookkeeping records in the office. This means an attacker could potentially have a direct path to those same bookkeeping records.

Additionally, any device that is accessible remotely can provide a starting point for an attacker to access every other device on your network. Once an attacker has the smallest foothold in your network, expanding the attack to other (or even all) devices on the network is exponentially easier.

Action Plan

- Ensure that remote access is only provided when absolutely necessary. If a team member does not need remote access, they should not have any trace of this access.

- Ensure that remote access technology requires multi-factor authentication (MFA). For example, most VPNs do not require MFA by default. This is a major risk. You need to ensure your IT team adds an MFA mechanism for all remote access options. If MFA cannot be enabled, do not use that remote access technology.

- Ensure you have a subscription for software updates for every type of remote access software employed.

- Ensure you have a person or team professionally managing the remote access technology and updates.

- Use additional safeguards with devices that can be accessed remotely. Since they are on a network with other devices, it means all of those devices. The additional safeguards could include SOC services and zero-trust controls. SOC services will help by carefully monitoring the device for signs of compromise. Zero-trust controls are a technical mechanism of locking down parts of systems in a way that trusts absolutely nothing unless they were specifically tagged as safe in advance. For example, with zero-trust controls on your laptop, Google Chrome or Microsoft Edge would not run unless they were explicitly marked as safe in advance.

❑ My organization uses wireless networking.

Business Risk

If an attacker gains access to your wireless network, they can potentially reach every other device connected to your network. This means a criminal parked on a public street near a business location could access your internal network as if they were sitting at a desk inside your office. All they need is a crack in the wireless network security to let them in. This can put your entire business at risk.

Action Plan

- Use business-grade wireless access points. Consumer-grade devices simply don't provide the level of security necessary for business environments. The risk of using these is too great, so all access points should be professional, enterprise-grade equipment.

- All wireless access points should have active subscriptions for software updates. It's essential that they are always running the latest software versions with up-to-date security patches. Regular updates close vulnerabilities that attackers could exploit.

- Ensure you have a person or team professionally managing the wireless network and updates.

- Design your wireless network with a security-first strategy:

  > Administrative controls: Don't post wireless passwords on bulletin boards or in easily accessible places. Ideally, team members shouldn't even know the password, and only the IT team joins devices to the corporate wireless networks.

  > Technical controls: Configure your wireless networks so that only preregistered devices can connect. This means that even if an attacker manages to obtain your wireless passwords, they wouldn't be able to connect unless their device is explicitly authorized by your IT team.

INCIDENT REPORT

Criminals have parked their cars outside of targets for many years. The same wireless technology that makes it convenient for your team to work while roaming around your buildings also provides opportunities for hackers. The tools and equipment needed for wireless hacking are commodities today. The Russian hacking group APT28 has successfully used neighboring wireless networks, so they no longer need to drive to a target. Imagine this: Now your company can be at risk if the businesses with property next to yours is compromised (Greenberg 2024).

CLOUD PRODUCTIVITY SUITE RISKS

❑ My organization uses a productivity suite, like Microsoft 365 or Google Workspace.

Business Risk

These tools are essential for maximizing efficiency, and we're in an era where knowledge workers have never been more empowered. There's simply no substitute—your business needs a productivity suite to stay competitive.

The risks associated with productivity suites are significant. Even if a team member with the lowest level of access uses a weak password for their Microsoft 365 account, it could jeopardize the entire organization. Remember, an attacker only needs a single point of entry.

94% of organizations were victims of phishing attacks in 2023.

(Egress Software Technologies Inc. 2024)

If, for example, a janitor's email password is a simple, guessable term like a child's name or birthday, a cybercriminal could gain access. That mailbox can then be used to send targeted phishing emails to other team members, including mid-level and senior management.

Additionally, internal security within these platforms is often less robust than it should be. Employees, even those with limited access, may still have access to more sensitive company data than intended. It's not uncommon for businesses to leave general company data widely accessible. Managers often unintentionally store confidential information in locations that aren't adequately secured.

Action Plan

- Enforce MFA. Every team member using the productivity suite must be enrolled in MFA—no exceptions. The risk level is too high to ignore this crucial security step.

- Leverage a professional Security Operations Center (SOC). A SOC should actively monitor your productivity suite for signs of compromise. For example, if someone signs into the CEO's mailbox from an unexpected location, such as another country, alerts should be generated and acted upon immediately.

- Implement redundant monitoring systems. In addition to the SOC, your regular IT team should have a separate alerting system to identify unusual behavior. The goal is to have multiple layers of oversight on this mission-critical system.

- Apply the principle of least privilege for data access. Data stored within your productivity suite should be segmented by access level. Avoid a single, catch-all data repository. Instead, categorize data and assign access based on need. When appropriate, this should be on a temporary basis. For example, during a project, ten team members may need full access to project data. Once the project is completed, perhaps reduce access to just two senior members when the data is archived.

INCIDENT REPORT

An Illinois woman, Erika DeMask, lost nearly $1 million in a pig butchering fraud. In her case, the criminal was playing the role of a potential romantic partner who asked her to make investments. The scam artist had promised to double her investment, although the requests appeared to have mixed language, such as needing money to repair equipment. DeMask took out a home equity loan in order to send more funds to the criminal. She was forced to sell her home as a result. She only has $400 to her name now (Pistone and Knowles 2024).

- Secure email communication. Email remains the most popular attack surface. To reduce risk, deploy multiple layers of email security. Traditional spam filtering is not enough today. It often does not even handle basic threats. Consider a next-generation phishing filter that's more aggressive. This blocks more sophisticated phishing attempts. It also continuously reminds team members of the ever-present threat landscape.

- Data backup is a last resort solution to a cybersecurity incident. That being said, it is mission critical in case everything else fails. Your productivity suite should have an independent, third-party backup mechanism for all data. The data should be stored in such a way that it cannot be destroyed by a criminal, since this is common practice in ransom attacks.

SERVER RISKS

❑ My organization uses one or more servers.

Business Risk

Servers are the backbone of your business, providing mission-critical services such as identity services, data storage, and more. No matter what role a server plays, even the least significant one is a valuable attack surface for cybercriminals. If an attacker gains control of your smallest, least important server, they can often leverage it to access everything else in your organization. Remember, an initial breach is just the starting point—not the end goal. In short, the risk is enormous.

Action Plan

- Implement the principle of least privilege. No one should have access to a server unless it's absolutely necessary for their routine work tasks. Team members who require access should be limited strictly to the services or data they need—and nothing more. Work closely with your IT team to audit access levels and remove any unnecessary permissions. This reduces potential entry points for attackers.

- Use an SOC. While IT managers may handle server maintenance, their role is not to constantly monitor servers for signs of compromise. You need security professionals actively watching every server 24/7. A dedicated SOC team excels in this area, using proprietary software to detect anomalies. If something suspicious is detected, they can jump in and investigate immediately, at any time of the day or night. If your business relies on servers, having an SOC team is non-negotiable.

- Implement zero-trust controls. If possible, you should deploy zero-trust security controls in your environment. This involves multiple IT teams assessing each server and defining strict access rules and communication pathways between servers and devices. By default, servers are overly trusting, which creates vulnerabilities. Implementing zero-trust means explicitly carving out what's allowed, effectively reducing the risk of malicious interactions.

- Ensure physical security is in place if one or more servers are located physically at your business locations. Physical access to servers is just as critical as digital security. Only team members with a legitimate need should have physical access to your servers. Ensure all servers are housed in locked racks or cabinets, which are

The median ransomware demand was $650,000 in 2023.

(Palo Alto Networks, 2024)

themselves located in locked, secure rooms. This applies to all critical network equipment, not just servers. If one or more servers are located physically at a third-party location, such as a data center, the third party should be ensuring physical security. They should also be virtually isolating your servers from other servers, so if another customer's server is breached, the attacker cannot easily access your servers as well.

- Ensure the IT team that set up servers in a public cloud (i.e. Microsoft Azure, Google Cloud, or Amazon Web Services) carefully configured the networking, firewall, and other security to limit access to the bare minimum. The default configuration for cloud servers is often publicly accessible from all other devices on the internet, which is an extraordinary risk.

- Ensure you have a subscription for software updates for all software installed on all servers.

- Ensure you have a person or team professionally managing the servers and keeping everything up to date.

- Data backup cannot be your primary plan for a cybersecurity incident, but it is your best measure of last resort. Servers should have data backed up and stored in multiple locations, so in the event one copy of the data is destroyed by a criminal you have multiple paths to recovery.

ENDPOINT RISKS

❑ My organization uses workstations and/or laptops.

Business Risk

Your team uses their computers to access confidential company data and perform critical business actions. While there's always some risk from bad actors within your organization, the far greater risk comes from external criminals or malicious software gaining access to a team member's computer. If that happens, the attacker can impersonate the user and gain access to everything that the user has permission to access.

Consider this: if your highest-ranking executive has access to all company files, then any malicious software or cybercriminal with control over that person's computer can access those same files. This could be the single greatest business risk your organization faces. In fact, there have been cases where a single click on a malicious link in an email led to the complete destruction of a company's data. The risk could not be clearer.

Action Plan

- Ensure professional management of all computers. Every computer in your organization must be professionally managed. This means assigning a team to keep software on all devices up to date. Even in large companies, a single unpatched device has led to catastrophic data breaches. Regular updates are critical for closing security gaps.

- Leverage an SOC. Your organization should have a SOC actively monitoring all computers for signs of compromise. If any device begins to show indicators of a breach, the SOC team can take immediate action. This is not generally a task for your internal IT team, who are typically stretched thin with other responsibilities. It is also rare that an internal IT team is truly 24/7/365. A dedicated SOC has the expertise and resources to respond swiftly and effectively.

- Implement zero-trust controls on all computers. Whenever possible, apply zero-trust controls to your computers, just as you would for servers. This means adding an extra layer of security on top of the standard IT setup. Ensure that only white-listed applications and authorized communications are allowed on these devices. This significantly reduces the attack surface for potential threats.

"In over 90% of cases where attacks progressed to ransom stage, the attacker had leveraged unmanaged devices in the network."

(Microsoft Corporation 2024)

- Provide security awareness training for team members. Your team members are your first line of defense. They should receive regular training to recognize potential threats and understand how to respond. Establish clear processes so that when something seems out of the ordinary, team members can report it immediately to a security professional for further review.

- Give laptops used outside the office a Secure Access Service Edge (SASE) solution. This would keep all laptops behind a virtual firewall, and they would be protected as if they were physically in the office.

- Backup is not always critical for workstations and laptops, unless they have data not synchronized to a cloud data system. Certain team members may have unique software that stores important data in locations that do not have any connection to cloud storage. For these computers, you should implement full computer data backup.

INCIDENT REPORT

In September 2023, Johnson Controls had a cyber incident. Criminals claimed to have 27 terabytes of their data and ransomed it back for $51 million. Large portions of their IT infrastructure were shutdown, which affected both internal teams and customers. While many of the incident details were not publicly disclosed, the total cost of the incident was reportedly $27 million (Toulas 2024).

This incident report is a great example, but keep in mind that typical targets for cybercriminals are much smaller. For example, there was a two-person accounting firm hit with a ransomware attack. The criminals had access to this firm's bookkeeping data of its own resources. The ransom was equal to the balance of their operating checking account. Unfortunately, they had no choice but to give the criminals every dollar or go out of business.

❑ My organization uses mobile devices.

Business Risk

This is obvious when your organization provides cell phones, tablets, or mobile/handheld computers. It may be less obvious if employees are using email on their personal cell phones. Employees may not ask for written permission in advance. Devices owned by the organization and those owned personally count for our purposes.

Just like workstations and laptops, mobile devices pose significant risks to your organization. What makes mobile devices unique is that, in many cases, they aren't corporate owned. Employees often use their personal cell phones to access company resources, such as email accounts. Some businesses also intentionally or accidentally allow employees to connect their personal mobile devices to corporate wireless networks. This creates a unique set of challenges because businesses may not want to incur the cost of providing company-owned devices solely for marginally better security.

In these cases, the risks are significant, but the controls you can apply directly to the devices are limited. Instead, you'll need to rely on security measures at other levels.

Action Plan

- Provide security awareness training for team members. Just as with workstations and laptops, it's crucial to train team members to recognize potentially malicious activity on their mobile devices. They should also be aware of the proper process for reporting any suspicious behavior to your security professionals. Awareness is your first line of defense.

- Provide team members with a clear process for reporting lost or stolen mobile devices so that your IT team can take appropriate steps to reduce any cybersecurity risks.

- Leverage an SOC. Your SOC should monitor all systems and data that employees access via mobile devices. For most small and medium-sized businesses, this primarily involves your productivity suite (e.g., Microsoft 365 or Google Workspace). If your SOC is already keeping an eye on this environment for suspicious activity, you may not need additional monitoring specifically for mobile device access.

- Enforce MFA. As with your productivity suite, all team members should have MFA enabled on their accounts accessible from mobile devices. This extra layer of security significantly reduces the risk of unauthorized access, even if an attacker gains access to an employee's mobile device.

❑ My organization uses printers, industrial or commercial equipment with networking, or other Internet of Things (IoT) devices. These could be networked devices such as environmental monitors or sensors.

Business Risk

As we've discussed, if a criminal takes control of any device on your network, it can potentially give them access to other—or even all—devices connected to the network. It might be hard to imagine how an old printer could compromise your entire company's confidential data, but that's the reality. You might think that because the printer is physically located in your secure office, and only trusted employees have access, there's no risk. But that's not true.

Every day, your employees interact with potentially malicious websites. A single compromised webpage could deploy malware designed to scan your network for vulnerable devices, like your old printer. If that printer is running outdated software, that software's security vulnerabilities could be exploited to give criminals a bridge into your company's systems. This is a serious risk—a ten-year-old printer could literally put your entire business in jeopardy.

Action Plan

- Ensure professional management of all networked devices, including printers, industrial/commercial equipment, and other IoT devices. Your printers and other IoT devices should be professionally managed by an IT team responsible for keeping their software up-to-date. This includes regular software patching for security vulnerabilities.

- Implement zero-trust controls across your network. If you can implement zero-trust controls on all devices, including printers, you can significantly reduce the attack surface. For example, if all workstations and laptops are configured with zero-trust policies, any unexpected communication from a printer will be blocked unless it is explicitly approved. Here's how this works: If a printer that has only ever accepted print jobs for years suddenly attempts to connect to your servers, zero-trust controls will automatically disregard these unauthorized requests. This prevents malicious software from using a compromised device to infiltrate other systems on your network.

- Employ network segmentation. When designing networks, your IT team can isolate or segment different parts of your network. For example, they can design it so your warehouse computers can only communicate with warehouse printers. This means if a warehouse printer becomes compromised, the attacker may not be able to access finance computers. This does not totally eliminate risk, but it does reduce potential harm in the event of a security incident.

INCIDENT REPORT

In November 2023, the Municipal Water Authority of Aliquippa, Pennsylvania, was compromised by the CyberAv3ngers—a hacking group affiliated with the Iranian Revolutionary Guard Corps. They targeted the water pressure monitoring system at a remote pumping station by exploiting a publicly exposed Unitronics Vision Series programmable logic controller (PLC).

In January 2024, the small Texas towns of Muleshoe and Abernathy were targeted by the Cyber Army of Russia Reborn, a hacking group allegedly linked to Russia's military intelligence. The group claimed responsibility for manipulating the human-machine interfaces (HMIs) at local water facilities, which led to the overflow of water storage tanks and caused minor, temporary disruptions to operations in Muleshoe.

There are an estimated 145,000 easy-to-compromise industrial control systems online today. As with so many cyber risks, it is not a matter of if they will be attacked, but when and what the consequences will be (Censys Research Team 2024).

Small businesses and even households are being targeted by criminals using newer IoT devices, such as smart home tools. Smart TVs, smart refrigerators, smart thermostats, and even smart fish tanks are all being used to break into computers used by the smallest of businesses. While public utilities being attacked makes headlines, when a small real estate brokerage goes out of business after a successful ransomware attack, almost no one notices.

HUMAN RISKS

❑ My organization has employees, individual contractors, or outsourced labor.

Business Risk

Humans are often the weakest link in cybersecurity. Humans have the keys to all the organization's critical assets. You must give humans access to some computers, data, systems, etc., for work to happen. Unfortunately, anyone with any level of access can provide that same access to anyone else in the world with the click on a single link. They only need to be fooled for a moment.

They can also fall victim to any number of confidence scams, such as wiring funds to criminals instead of legitimate vendors. They can direct deposit payroll to foreign accounts. They can send your vendor payment for ABC Supply to ABCD Supply without a second glance.

When your CEO asks for $2,000 in gift cards for clients, someone on your team may actually run to the store—we've seen PhDs and executives fall for these kinds of scams.

Keep in mind that one of the largest risks for individuals today is a category of pig butchering cyber scams.

As a business owner, you have additional concerns to consider. When employees are under significant financial pressure, it also increases the risks of embezzlement from any accounts they have access to. Your family and other business stakeholders may also make decisions otherwise unthinkable if they personally fall victim to a pig butchering scam. You do not need to personally fall for a scam like this to be negatively impacted.

Suffice it to say, the business risks here are enormous.

Action Plan

- Everyone with any access to any system needs cybersecurity awareness training.
- Everyone with any access to any system needs to know how to report suspicious activity.
- Everyone with any access to any system needs to know how to ask an expert if an email, text, or other message is safe or malicious.
- Everyone with any access to any system should be periodically tested with phish simulation emails. This will help you determine who may need additional training.

- Your business should have strong administrative controls, especially for finance. If one person falls for a confidence scam, simple policies could prevent an incident. For example, perhaps all wire transfers should require two functional approvals. For example, perhaps some changes need a face-to-face handoff, like employee direct deposit changes. There are some actions cybercriminals cannot perform. So, administrative controls can stop them before a loss occurs by requiring these actions.

SOFTWARE/ APPLICATION RISKS

❑ My organization uses web-based applications.

Business Risk

One of the greatest risks associated with using web-based applications is that a third party is responsible for protecting the confidentiality of your business data. This leaves many critical factors outside your direct control. Additionally, because these applications are designed to be accessible from any device connected to the internet, the attack surface is vast.

To better understand the business risks, it's essential to create a list of all the web-based applications your organization uses and assess the risks associated with each. For example, if an application is only used for booking travel, the risk might be relatively low. However, if another application stores sensitive health records, the risk level is extremely high.

Action Plan

- Create a risk profile for each web-based application. Start by categorizing each application based on the sensitivity of the data it handles. Understand which applications pose high risks and which are low-risk. This will guide your efforts in securing them, as web-based applications usually require tailored security measures rather than a one-size-fits-all approach.

- Implement Single Sign-On (SSO) wherever possible. Using SSO technology allows you to control access to all web-based applications from a single security console. For instance, if your applications are linked to Microsoft 365 identities, disabling an employee's Microsoft 365 account will instantly cut off their access to all connected applications. Without SSO, IT managers might need to manually revoke access from multiple systems. This increases the risk of ex-employees retaining access to some systems indefinitely. During audits, we often find that former employees still have access to systems long after their departure, especially in larger or more complex organizations.

- Enforce MFA for all accounts. As discussed earlier, enabling MFA on all web-based applications is crucial. While it doesn't eliminate all risk, it significantly reduces the chances of a bad actor successfully logging in to a team member's account.

- Limit access to mission-critical applications. For your most critical web-based applications, consider working with the vendor to restrict access based on location. For example, you can configure the system to only allow access from your company's office locations. While this may seem inconvenient for remote employees, you can implement solutions like VPNs to route their connections through the office network. This kind of technology lets remote team members appear as if they are in the office when accessing third-party web applications. This approach drastically reduces the number of devices in the world that can attempt to log in to your most sensitive applications, making it much harder for attackers to gain entry.

- Select vendors with robust security operations. For proprietary web-based applications, it is often challenging to have your own Security Operations Center (SOC) monitoring activity. However, when choosing vendors, prioritize those with mature security practices. Look beyond the application's features and evaluate the security infrastructure in place. Ensure the vendor offers granular access control, alerts for unusual activity, and other proactive security features.

- Web-based applications generally have some level of data backup provided by the software developer. The trouble here is that this is outside of your control. The developer may not have any significant liability or consequences in the event of data loss, but you probably do. You should utilize an independent, third-party backup solution for all web-based applications when available. Some web-based applications do not have a built-in mechanism for allowing independent backup. In those cases, you may be able to request periodic or automated data exports that can be backed up with your other company data.

❑ My organization uses third-party software.

Every organization relies on third-party software. It is simply a necessity for basic business operations. Third-party software is any application your business uses that wasn't made by your company. For example, Microsoft Word, QuickBooks, and Calendly are all third-party software. They are created and maintained by other companies.

This is slightly different than just web-based applications, although there is crossover. Any web-based application not written by your organization is third-party software. Third-party software can also be run directly on your company's computers, so there is a different risk profile.

Business Risks

The challenge lies in trusting that these third-party providers have secured their software well enough to prevent it from becoming a conduit for attacks on your organization.

Just like any device on your network can serve as an entry point for a criminal to access other devices, one compromised piece of software can be used to take control of a device and, potentially, your entire network. I'm sure you're seeing a pattern here: the business risk is enormous. A single vulnerability in one piece of software could potentially bring down your entire business.

Action Plan

- Maintain support agreements for ongoing security patches. Every piece of software in use within your organization must have a support agreement with the software developer to ensure access to timely security patches. If a vulnerability is discovered, you should be entitled to receive a new version that addresses the issue. If a software vendor cannot provide this, your business should not use their product. Full stop.

- Monitor and apply software updates regularly. It is not enough to simply have an agreement for updates. You need an IT manager or team dedicated to monitoring the software across all your systems to ensure everyone is using the most up-to-date versions. In some cases, updates can be automated, but in other cases, it may require manual checks. Someone on your IT team may need to set up a calendar reminder every 15 to 30 days to check with vendors for new updates, then audit all software installations company-wide. If any devices are running outdated versions, your IT team will need to coordinate with team members to get systems updated promptly. It is absolutely crucial for your business's security. These are security basics.

- Leverage an SOC. Just as with other devices and systems, any device running third-party software should be monitored by an SOC team for signs of compromise. Continuous monitoring ensures that any suspicious activity is detected and addressed immediately.

- Implement zero-trust controls on devices running third-party software. Ideally, all computers running third-party software should have zero-trust controls in place. If a piece of third-party software is compromised, zero-trust controls may be able to prevent it from being used to attack or gain access to other devices on your network.

- All data and files necessary for the third-party software needs to be backed up using a comprehensive data backup system. This may already be covered by your server backup program. You may also need to consider backing up full workstations and laptops if the installation of such software is complex, time-consuming, or some of the data is stored directly on the computer.

INCIDENT REPORT

GitHub is grappling with a large-scale attack involving millions of malicious source code repositories aimed at stealing passwords and cryptocurrency. These repositories are clones of legitimate ones, created through an automated process that embeds malware concealed under multiple layers of obfuscation. Retaining the original repository names, they are difficult to detect and have been unwittingly forked (copied by other software developers), further spreading the attack.

Despite GitHub's efforts to remove malicious repositories, many evade detection due to advanced automation or manual uploads. Researchers estimate that over 100,000 repositories have been impacted, with millions of malicious forks uploaded before removal.

GitHub has responded by emphasizing its use of machine learning and manual reviews to detect and remove abusive content. The platform also encourages community reporting as it adapts to evolving adversarial tactics. This incident highlights the challenges of securing large-scale collaborative platforms (Goodin 2024).

In this case, the news headline is about GitHub, but the real victims here are the hundreds of millions of businesses and individuals that rely on source code being maintained on GitHub. Even if you've never heard of GitHub, there's software running on your computer right now that came from it. Without digging too deep, just recognize that programmers around the world need to collaborate, even on small pieces of source code. This is the most popular place for them to work together. Source code from GitHub is used in nearly every modern website and application.

❑ My organization uses custom applications.

This would be any software you paid a developer to create for your organization. Keep in mind that complex Excel spreadsheets or analytic dashboards often have custom source code in the background. You may need to ask team members or third parties if any custom source code is in use.

Business Risk

While it might seem that custom-built applications offer greater control and reduced risk compared to third-party software, the reality is often quite different. In fact, custom applications can introduce even more complex security challenges.

Developers, whether internal or external, often recommend custom solutions based on the initial development time frame, without fully considering the ongoing maintenance costs. The risk here is that without a dedicated team to manage and update the application continuously, vulnerabilities can go undetected for extended periods.

Custom applications can have more vulnerabilities than off-the-shelf software. Unlike third-party software that often benefits from continuous updates and security patches provided by the vendor, custom solutions rely entirely on your development team for ongoing security. If you don't have a team monitoring and patching vulnerabilities daily, your business may be exposed to significant risks.

Custom source code should be recompiled periodically to keep up with updates in development languages and tools. This is essential to ensure they are utilizing the latest software libraries and methodologies. However, this process can be costly and time-consuming, especially if it's not accounted for in the initial budget. Failing to update custom applications could leave your organization exposed to outdated, insecure source code.

Action Plan

- Where possible, move to professionally managed third-party software. When possible, look at moving your business workflows to third-party software that has full-time developers mitigating the security risks in the source code.

- Many applications have the ability to run custom routines that will allow your team to have the custom experience they want with lower risk than fully custom software. If third-party software can't fully replace custom software, consider custom add-ons to extend its functionality. This will keep the security of the core system high and limit your attack surface to a much smaller element.

- If custom software must be used, you need two contracts with different vendors. One should be with professional developers to maintain your application, unless you have internal developers. This team must update the custom software periodically and definitely within 15 to 30 days of a new security vulnerability being discovered. You should have a separate contract with a professional security firm to test your custom application for vulnerabilities on a recurring basis and provide guidance for minimizing the risks posed by the application.

- Your custom application source code and data need to have solid backups with long version histories. This may already be covered by your server backup program.

DATA STORAGE AND MANAGEMENT RISKS

❑ My organization uses one or more databases.

If you have third-party or custom applications, you may need to check with the developers to understand if the data is stored in a database. For any line of business application, you should ask the developer if any database is located on your servers, workstations, or laptops. For web-based applications, your business data is most likely in a database on the software developer's servers.

Business Risk

Databases are the primary technology used for storing critical business records. It should be clear that these are prime targets for cyberattacks. Taking over one database could be enough to bring many businesses to their knees, which makes for quick ransom negotiations with cybercriminals. Business data, especially in databases, is an incredibly big target and is therefore high risk.

Average time to identify and contain a data breach is 292 days.

(IBM 2024a)

Action Plan

- Sensitive data should be encrypted. The encryption keys should be stored securely and separately from the encrypted data. In some cases, you may be able to have multiple layers of encryption.

- Databases should have a robust backup strategy designed for rapid recovery. Depending on the number of transactions your business has in a day, you may want backups every six hours or even every five minutes.

- For more complex environments, perhaps you want multiple database servers across different geographical regions, all synchronizing data in real-time. Talk with your IT team about listing your databases, their criticality for business operations, and plans for recovery in the event of a cyberattack or other system failure.

❑ My organization uses cloud storage.

For example, Microsoft OneDrive, Teams, Dropbox, Google Drive, and Box. Most likely, you have team members using cloud storage, regardless of whether or not it is company-approved. Your IT manager is also most likely storing some data, such as backups, in the cloud.

Business Risk

Business data has multiple risks to consider. First, how will your organization be affected if everyone lost access to all of your business data? This is what happens if a criminal successfully compromises a data storage system, encrypts it, and then asks for a large ransom to return your data to you. Ransomware attacks are extremely popular, ever since the attack techniques matured 20 years ago. International criminal organizations are making incredible fortunes from daily ransomware attacks on businesses. They have no ethics at all. Hospitals—even children's hospitals—are prime targets since they generally pay ransoms quickly.

61% of organizations reported breaches of their public cloud environments within the last year.

(Cybersecurity Insiders, Inc. 2024)

Another risk: How will your organization be affected if a criminal copied your confidential company data and then released it to the general public? Maybe the confidential records would just create embarrassment. More likely, the confidential data would create a series of legal challenges, as your customers, employees, vendors, and others line up to collect damages from the remains of your organization. If the data you stored was only yours, the risk may be low. But that is not realistic, since your organization likely holds confidential data related to many others.

Some of the most significant data breaches occur when patient data is compromised, since so many extremely personal details may be publicly disclosed. For this reason, HIPAA violations have steep penalties to protect the public. Financial and HR records, such as those containing Social Security Numbers, are also high-profile incidents due to the impact on the general public.

Action Plan

- Ensure an SOC team is carefully monitoring all data storage locations. If you are using cloud storage that cannot be monitored by your SOC team, consider moving to a cloud storage service that can integrate with your security team. Your SOC needs to be able to watch for indicators of compromise across all of your critical systems in order to best protect your organization against all threats. Having any element, such as one cloud storage system, outside of their purview can put your organization at risk.

- While your cloud storage service has some redundancy or backup built-in, you cannot risk your organization to their internal backup systems. You need to implement multiple layers of backup that stores data in multiple locations so if one or two locations are compromised, you still have a road to recovery available. Think of this as having multiple fire extinguishers available for the same fire—it is good planning.

- Use the principle of least privilege for access control. We discussed this earlier, but in a nutshell, if every team member only has access to the data required for their day-to-day tasks, you significantly reduce the risk to your organization. If most people have access to most data, then a criminal just needs to compromise any random account to attack your organization. If the criminal has the added challenge of needing to compromise a small number of specific accounts to get at the data they want, it may discourage them from continuing the pursuit. This is especially true if those with the most access are well trained to recognize and avoid phishing and social engineering attacks.

❑ My organization uses backups.

Business Risk

Unlike other sections we've discussed, the business risk here is a bit different. In this case, backups reduce your business risk — but only if they're set up and maintained correctly. The real danger is if you assume your business is safe because you have backups, only to discover they're unusable when you actually need them. That is like driving a car with faulty airbags.

While most IT managers implement some form of backup system, many do not actively monitor, maintain, or test those systems. Worse yet, even a fully functional backup system might fail during a cyberattack. This is because one of the first things cybercriminals often do is destroy or corrupt your backup data. If your backups are compromised, your business could face catastrophic data loss, putting its very survival in jeopardy.

Action Plan

- Establish a backup system for every data location. If you don't already have a backup system in place for every location where your business stores data, this is your top priority.

- Start by creating a list of all systems where your data is stored, and label each one according to the significance of the data it holds. Then verify that a backup solution is in place for each of these systems.

- Ensure that each backup system is actively monitored, regularly maintained, and periodically tested.

- Work with your IT manager to confirm that backups are not only running but are also being tested for restorability. It's crucial that any changes to your data environment are reflected in your backup strategy.

- Implement the 3-2-1-1-0 backup strategy. Speak with your IT manager to determine which of your current backup systems conform to this 3-2-1-1-0 strategy. For any systems that don't, explore options to align them with this best practice.

- Ideally, every backup system should follow the 3-2-1-1-0 strategy:

  > 3: Keep three or more copies of your data (one original and two backups).

  > 2: Store the backups on at least two different types of media (e.g., hard drives and cloud storage).

  > 1: Keep at least one copy off-site in secure storage or

the cloud.

> 1: Ensure that one copy is immutable, meaning it cannot be altered or deleted, even by your IT manager. This is critical for cybersecurity, as it prevents attackers from easily destroying your backups.

> 0: Confirm that your backups are error-free. The IT team should conduct regular test restores to verify that there are zero errors.

PHYSICAL RISKS

❑ My organization has a location with computer hardware.

Business Risk

There's an old legal expression that possession is nine-tenths of the law. For cybersecurity, you should consider that physical access to a device for a criminal, even for thirty seconds, can easily be full practical "ownership"—a 100% breach.

The techniques change from time to time, but here's a simple example: Let's say a criminal inserts a USB thumb drive into a receptionist's computer. This computer may actually install malicious software instantly and automatically. This would happen without any team member knowing. This happens in the background, and it happens all the time.

As you already know, a small amount of malicious software on one receptionist's computer can become malicious software on all computers in minutes or hours. There is no higher risk for your business than your computers running malicious software, other than perhaps a fire, flood, or natural disaster.

Action Plan

- Higher-level equipment, such as servers, network switches, etc. should be in locked cabinets and also ideally within locked rooms. This way if maintenance has access to the room or closet, they do not also have direct access to the equipment. Only IT or senior leadership should have such access.

- Lower-level equipment, such as workstations, laptops, etc. should have a few types of protections:

    - All data on devices should be encrypted. If any data is not encrypted and the device is taken, your confidential data can become public. Criminals can also use this data to craft new, custom attacks on team members to further compromise the organization or your customers and vendors.

    - For team member devices, your IT team should configure policies to automatically lock computer screens after a few minutes of inactivity. This will reduce the risk of unauthorized access if they walk away from a computer without locking it first.

    - Depending on the sensitivity of data your team members look at, you may consider privacy screens. This is especially important for laptops that are used in public settings. This reduces the risk that unauthorized people can see sensitive information unless they are standing directly in front of a monitor.

    - Equipment in public settings should be tethered to

furniture to prevent easy removal.

> Your IT team should use security software to limit USB port access to trusted devices, if possible. This would reduce the risk posed by one of the most common types of attack: malicious thumb drives.

❑ My organization uses video surveillance, access control (i.e., keyless door locks), intercoms, alarms, etc.

Business Risk

As we've previously discussed, small computing devices, such as security cameras, are frequent targets of cyberattacks. These devices pose a unique challenge for organizations because, while they're susceptible to the same kinds of attacks as regular computers, they often lack the power to run sophisticated security software.

Imagine this scenario: a single insecure video camera on your network could potentially compromise your entire organization, just as any other hacked device could. The risk is real, but it can be managed with a thoughtful approach.

Action Plan

- Implement network segmentation. Speak with your IT manager about segmenting these devices from the rest of your network. This means isolating devices like security cameras so they can only communicate with each other and not with your core servers or computers. For example, in the event a camera is compromised, an attacker would be limited to accessing only other cameras, greatly reducing the impact of a breach.

- Choose trustworthy manufacturers. Be vigilant when selecting manufacturers for your security devices. Years ago, the USA banned Hikvision surveillance systems due to national security concerns (Federal Communications Commission 2022). Keep in mind that over 600,000 Hikvision cameras were installed throughout the country before this ban (Hillman 2022). Counterfeit Cisco network switches were installed around the world possibly years before anyone caught on to the fraud. Your supply chain matters for everything your organization uses, including all technology devices. To reduce risks:

  - Select reputable manufacturers with a proven commitment to security.

  - Ensure the equipment is sourced directly from the manufacturer to prevent tampering during shipping.

  - Confirm that the manufacturer provides routine updates and patches for identified vulnerabilities. Prompt updates are essential for keeping your organization secure.

- Have professional management and monitoring. Treat these devices like any other critical part of your IT infrastructure. Your IT team should actively monitor the status of all connected devices. If a camera hasn't received updates or patches in a timely manner, it should trigger an alert. This proactive approach can help prevent potential breaches before they occur.

- Use cloud-based storage for security footage. In physical security incidents, one of the first things criminals often attempt is to disable security cameras or compromise the devices recording the footage. To counter this, store your security camera footage in the cloud as quickly as possible. This ensures that even if attackers disable cameras or on-site recording equipment, you may still be able to recover some footage.

- Review your current system and compare it with other modern, integrated physical security solutions. Not all surveillance systems offer robust cloud storage options. Selecting a system with strong cloud capabilities can make a significant difference in protecting your business during a physical security event.

COMPLIANCE AND REGULATIONS

❑ My organization has cyber insurance, wants cyber insurance, has regulatory requirements, or must meet industry standards.

For example, think about GDPR, HIPAA, PCI, or FTC Safeguards. Don't skip ahead just yet. HIPAA extends to business associates of HIPAA-covered organizations. For example, if you provide any service to an organization in the health care space, you may have some HIPAA compliance requirements. Also, the FTC expanded the list of organizations required to comply with the FTC Safeguards. It now includes almost all businesses that interact with credit, financial, or asset services, even indirectly, such as real estate appraisers.

Business Risk

The volume of new compliance regulations and industry safeguards is increasing almost exponentially. Even if your industry hasn't been hit with new requirements yet, you should prepare for them to arrive soon. Additionally, even if regulatory bodies are slow to impose cybersecurity standards, your vendors or clients may still require them to maintain business relationships.

> **63% of underwriters ranked ransomware as the number one threat.**
>
> (Woodruff-Sawyer & Co., Inc. 2024)

For years, many Fortune 500 companies have refused to work with organizations that don't meet certain cybersecurity standards or lack cyber insurance. For some of our clients, meeting these requirements has been mandatory for years. Failing to meet these standards or being unable to renew cyber insurance could result in significant revenue loss if customers decide to take their business elsewhere.

The business risk here is substantial. Even if your customers don't have strict requirements today, a single cybersecurity incident that damages your reputation could cause customers to leave. While much of what we've discussed thus far has focused on technical controls, this section emphasizes the importance of administrative controls.

Handling compliance requirements isn't just about implementing cybersecurity measures; it's also about documenting them in detail. If your organization faces a cybersecurity incident, you need to demonstrate that you took all necessary precautions to avoid being deemed negligent. Failing to prove compliance could expose your business to penalties beyond the direct impact of a cyberattack.

If regulatory bodies impose fines, industry groups apply penalties, or stakeholders file lawsuits, having comprehensive documentation will be critical.

Action Plan

- Establish and document cybersecurity policies. Regulatory bodies and industry groups often require a set of documented policies, signed off by your team. These policies might include areas such as access control, data protection, incident response, and employee training. Ensure you have these policies in place and that they are regularly reviewed and updated.

    - > Acceptable Use Policy
    - > Data Confidentiality Policy
    - > Bring Your Own Device (BYOD) Policy
    - > Password Policy
    - > Mobile Device Policy
    - > Incident Response Policy
    - > Backup & Disaster Recovery Plan
    - > Remote Access Policy
    - > Security Awareness Policy

  - IT Asset Disposal Policy
  - Third-Party Access Policy
  - Removable Media Policy
  - User Termination Policy
  - Business Continuity Plan

- Maintain ongoing documentation of compliance. It's not enough to have written policies; your IT team must also be able to regularly document compliance with both your internal policies and the technical controls required by your industry or insurance providers. This documentation should include proof of regular updates, audits, and adherence to best practices.

Conduct third-party audits quarterly or at least annually. For compliance and insurance, a third-party auditor must review your cybersecurity measures at least once a year. This audit should review your policies, procedures, and technical controls. It should also run technical tests to find any security gaps. These audits not only help ensure compliance but also demonstrate due diligence in the event of a cybersecurity incident. Having these audits quarterly will improve your ability to successfully resolve a claim, since the insurance carrier will have ample, recent evidence of compliance with their underwriting requirements.

SUPPLY CHAIN AND THIRD-PARTY RISKS

❑ My organization has vendors, customers, or partners.

Business Risk

External parties pose a magnitude of risks to your business. They often have confidential—or at least non-public—information about your business. They decide how to protect that data. They decide which internal team members have access to the data. Aside from a nondisclosure agreement (NDA), you may not have any controls.

What would happen if a key vendor or customer had a data breach? If non-public data about your business was now public, how would your business be affected?

Now, let's consider a different scenario. One key supplier has a cyberattack. The criminal has taken their entire business offline. Now, they cannot provide the contracted goods or services. What would the impact be to your organization? How many hours, days, or weeks can you continue without that supplier? What would the impact be to your customers?

The more critical your organization's role in the supply chain, the less likely you can weather hours or days of downtime.

Many businesses, such as the Fortune 500, have detailed supply chain risk strategies in place. They started adding cybersecurity baselines in their supply chain agreements. Today, if you have a Fortune 500 customer, you may already have contract obligations to carry cyber insurance. Cyber insurance is not offered to businesses without a certain level of cybersecurity baselines. This means your customer knows you have at least basic safeguards in place. This gives them confidence to do business with you and trust you as part of their critical supply chain.

Now, think about the outside parties you work with. What danger would they put your business in if they did not take cybersecurity seriously?

Action Plan

- List the external parties that you rely on. Note the risks each party poses to your organization. Which only risk confidential data being public disclosed? Which pose operational risk if they went offline?

- For each: Think about how you can contractually insist they maintain basic cybersecurity protections.

- For each: Consider your plan if they have a security incident. Have an idea of how to respond to a data breach. Have an idea of how to respond to downtime.

Congratulations!

You have just taken a giant leap forward in securing your organization.

You now have a checklist of exact steps you will need to take. As you went through this section, hopefully you made notes of which steps you believe are most important.

Next up: decide which steps you want to take action on first, then decide who you need to speak with to take the next steps.

I hope this exercise has helped you think about your organization's risks.

CHAPTER SIX

BUILD ON A STRONG FOUNDATION

While these administrative and technical controls may look good, you should not build your cybersecurity program on sand. You must have a solid foundation from the leadership.

There are some critical pieces you need to have in place for your cybersecurity program to be successful:

- Leadership must set the tone. No one in your organization is going to take cybersecurity seriously unless the executives do. The executives and managers must agree to continuously communicate the importance of cybersecurity. When the leaders insist cybersecurity priority, the teams are much more likely to listen.

- IT team members must be empowered. No one in IT wants a security incident, but they need the authority and resources to make security a priority. Our IT firm requires that service requests, such as password resets, follow security-first workflows. You cannot simply call in and ask for a new password. The security-first organization has significant changes, which may make some workflows less convenient and can cause friction with team members. Of course, they are more convenient than handling a major cybersecurity incident.

- Third-party cybersecurity experts should be consulted. Even international conglomerates rely on third parties to assist with cybersecurity. It is important to have independent third parties involved for the best comprehensive and unbiased advice. They can also be helpful if your internal IT team needs assistance with implementation projects. Not all IT firms that manage technology can provide comprehensive cybersecurity programs. You may consider bringing in a cybersecurity-focused firm in addition to your regular IT firm. If necessary, perhaps move your technology management to a security-first firm.

- To make your cybersecurity program a reality, you need to schedule time on your calendar to act. My suggestion is to bring in a third-party to assess your cybersecurity every ninety days. This is ideal for many reasons, anyway, including meeting cyber insurance and regulatory compliance requirements. This also means you'll have a meeting at least once a quarter to see where your cybersecurity posture stands. From these meetings, you can move through the open items on your cybersecurity program to do list and ensure forward progress.

Please do not allow your organization to lose momentum. Cybersecurity programs take time and energy to implement. Don't waver. Take steps every quarter until your plan is rolled out completely.

CHAPTER SEVEN

TAKE THESE STEPS TODAY

If you are starting from scratch, or almost scratch, you may seem overwhelmed by the number of steps you'll need to take.

Let's avoid the overwhelm and hit the road running today.

These cybersecurity program components will be fastest to implement. You can send the emails or make the phone calls to get these started before the sun sets.

- Third-party security assessments. As we already discussed, you should have these for many reasons. The safety of your organization depends on these vital reports. You also need them for compliance with insurance, regulatory, and industry requirements. You can engage a company today and get it scheduled.

- SOC services. It would be difficult for most internal IT teams to provide this, so you will need a third-party provider. They will have a full team working 24/7/365 with operational authority to isolate computers posing serious risk to your business. They will provide reporting to either your IT team or your point-of-contact in leadership about any potential incidents. You can engage a company to provide these services today, although it may take a bit of time for them to implement full service.

- Security updates for most devices should be part of your basic IT management and support agreements with your IT firm or handled by your internal IT team. You can quickly ask them to provide a summary of which devices are getting regular updates and the frequency. With that summary in hand, you can compare it with this list to ensure nothing is being missed. If anything is being missed, ask your current IT team what will be required to add frequent updates for the missing devices.

    > Firewalls

    > Network Switches

    > Wireless Access Points (APs)

    > Servers

    > Workstations and Laptops

    > Printers

    > Any other devices connected to your networks

These three should give you some quick wins and kick start your progress to a more secure cyber future.

Once these are checked off, the next one I would personally recommend tackling is ensuring that every account for every system has MFA enabled. This may take some time and some auditing, but it will provide an excellent return on your time investment.

CHAPTER EIGHT

CYBER INSURANCE REQUIREMENTS

We discussed many controls in this book. Now let's look at this briefly from a different angle. If you're applying for a cyber insurance policy (or renewing one), here are some of the most common requirements. Most carriers will have significantly longer lists and many more administrative and technical controls, but this is a good place to start.

Credential Management

- Require unique and complex passwords.
- Prevent the reuse of previously used passwords across multiple services.

Physical Security

- Restrict physical access to protected information assets to authorized persons.
- Prevent tampering with organizational infrastructure by restricting physical access to secure areas.

> **Cloud account credentials alone make up 90% of for sale cloud assets on the dark web.**
>
> (IBM, 2024)

Employee Training

- Train all employees and third-party providers on approved security and privacy policies.
- Educate employees on the threat of insider threats, social engineering, and privacy policies.
- Update security training annually and evaluate employee understanding.

Access Control

- Approve and monitor new user access to data.
- Require authorization for information system access.
- Segregate data networks and minimize privileged account access.
- Require MFA for all sensitive systems and all external access.

- Have a formally documented process for immediate removal of access for separated employees.
- Set conditional access policies for Microsoft 365.
- Privileged Access Management (PAM) system for administrative access.
- Alerts for user access changes, such as new user account creation.

Data Protection

- Secure all access to sensitive data and enforce role-based access controls.
- Encrypt and authenticate data transmission methods.
- Eliminate unused functions, services, and protocols regularly.

Communication

- Enforce Sender Policy Framework (SPF), DomainKeys Identified Mail (DKIM), and Domain-based Message Authentication, Reporting, and Conformance (DMARC) policies on all inbound and outbound email.
- Mark all inbound messages from external sources clearly

with a warning.

Incident Response & Business Continuity

- Develop incident response plans and define authorized responses.

- Monitor for security incidents, respond promptly, and retain audit records.

- Make a formally documented Business Continuity Plan that is updated and tested annually.

Security Measures

- Implement endpoint detection and response technology to prevent malicious code from executing.

- Apply system updates promptly.

- Use encryption for backups and critical information, and perform test restores.

- Implement zero-trust controls for end-user devices.

- Employ a Security Operations Center (SOC) team to monitor systems for indicators of compromise.

- Limit wireless network access to specific devices (by MAC address) and disable broadcasting of network names.

Compliance and Governance

- Request third-party security assessments at least annually, including internal and external penetration tests.

- Promote ethics and compliance culture. Senior management must do this.

- Designate an Information Security Office (ISO).

- Enforce dual authorization for high-risk transactions and changes to controls.

- Conduct background checks for new employees and maintain third-party contracts with security requirements.

Vendor Management

- Have a formal Vendor Management Program.

- Have contractually obligated cybersecurity requirements for vendors.

- Have contractually obligated cyber and crime insurance

requirements for vendors.

This is just a starting point to ensure you can start the application process. Your specific insurer may require you to implement more controls either prior to coverage starting or during the first six month of your policy.

Cyber insurance carriers are looking for ways to deny claims since the dollar amounts for cyberattacks can be staggering. Exceeding your carrier's requirements is important. Documenting your compliance with these requirements so you can prove that you were in compliance every day of every year prior to an incident is even more critical.

If your carrier is able to demonstrate that your systems may have fallen out of compliance even an hour before a cyber incident, your claim may be denied.

Next, let's take a look at what happens if an incident occurs.

CHAPTER NINE

CRIMINALS HAVE ACCESS. WHAT NOW?

Safe driving cannot prevent all vehicle collisions. Regular doctor visits do not prevent all disease. Campfire safety education does not prevent all forest fires. Preparation does not guarantee the desired outcome.

Implementing a cybersecurity program is critical, but no program can prevent all attacks.

Here's a small list of governments whose agencies had recent attacks, despite having cybersecurity programs:

- NATO
- African Union
- Australia
- Belarus
- Cambodia
- Canada
- Czech Republic
- Denmark
- Ecuador
- El Salvador

- Germany
- India
- Indonesia
- Israel
- Japan
- Kuwait
- Mongolia
- New Zealand
- Norway
- Poland
- Russia
- South Korea
- Sri Lanka
- Switzerland
- Thailand
- Trinidad and Tobago
- Ukraine
- United Kingdom
- United States
- Bernalillo County, New Mexico
- City of Atlanta, Georgia
- City of Baltimore, Maryland
- City of Oakland, California
- City of Riviera Beach, Florida
- Los Angeles Unified School District (LAUSD)

Many companies have comprehensive cybersecurity programs, but still suffer attacks. For example:

- Adobe
- Capital One
- Deep Root Analytics
- eBay
- Equifax
- Exactis
- Facebook
- First American Financial Corp
- FriendFinder Networks
- Heartland Payment Systems
- Home Depot
- JPMorgan Chase
- LinkedIn
- Marriott International
- Microsoft
- MySpace
- Progress Software (MOVEit software)
- Real Estate Wealth Network
- River City Media
- Target
- Yahoo
- Zanga

The list of breached organizations is practically endless.

Your organization will not be the first or last to experience a cybersecurity incident.

It is unrealistic to expect that every single security hole can be closed, especially because new ones are discovered every day. No one can, regardless of funding, expertise, or resources.

This is why your organization needs an incident response plan.

These plans come in all shapes and sizes since every organization is different. In some cases, this can be written by one or two people in less than an hour. In other cases, this may require a dozen meetings and include the heads of twenty departments over the course of a year.

You may want to have multiple response plans based on the trigger or threat type. For example, one malicious email may not warrant the same response as your primary business operations software being unavailable.

Since no template is going to work for everyone, let us instead focus on the core questions that must be answered:

Preparation

- Who should be included in the incident response team? This expands well beyond just IT since serious incidents will impact business operations. Consider these roles: incident commander (to oversee the entire response), IT, legal, compliance, communications (internal and external), HR, and other management. Hour to hour, day to day, there are likely more business operations decisions to make and execute than IT ones. Preparation is key.

- What logging and auditing do we want implemented before an incident, so the incident investigations can be performed?

- How will incident activities and tasks be tracked? Keep in mind that your primary system(s) may not be available.

- How will the incident response team communicate? Keep in mind that your primary communication channels may be compromised or unavailable.

"Fifty percent of small to medium-sized businesses (SMB) have been the victims of cyber attack and over 60% of those attacked go out of business."

Dr. Jane LeClair
Chief Operating Officer
National Cybersecurity Institute

(Bressler and Bressler 2019)

- What tools and software should be implemented before an incident? For example, there should be centralized security/event logging, endpoint detection software, etc.

- Which consultants or firms should be engaged or retained prior to an incident?

- In terms of communication, who needs to be notified, and when? In some cases, leadership and managers may need to be notified quickly, if the scope expands beyond a minor breach. In other cases, perhaps no one needs to be notified, other than the incident response team, until after the incident is resolved. In all cases, it is ideal to plan your communication in advance to reduce time investment for these decisions during the incident.

- Are there legal or regulatory considerations? In some situations, depending on the severity of the incident, you may only have hours or days to make legally required notifications to regulatory agencies. You may also have contractual requirements to notify your insurance carrier(s) or certain suppliers and customers.

- For cyber insurance requirements, you also need to understand at what point you must reach out to your insurer. Many carriers have strict requirements for incident handling if a claim will be filed. In fact, you may need to draft your incident response plans based on your specific cyber insurer requirements to guarantee coverage.

Identification

- What will trigger this response plan? For example, external or internal reports of unusual or suspicious activity would be typical triggers. Think through the first possible indicators of a possible incident so you can train your team on what to look for.

- If there is a possible incident, what will be checked to determine if there is, in fact, a real incident? How will scope be determined? For this section, you likely want your IT team or a cybersecurity specialist to draft a list of Indicators of Compromise (IoC).

- What logs can be reviewed to investigate the incident?

Containment

- What can be done quickly to contain the compromised systems? Will containment improve the situation or make

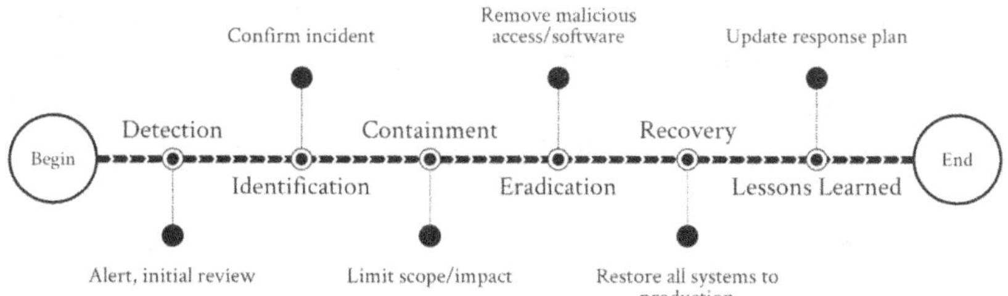

the situation more challenging? You may want to work with your IT team or cybersecurity consultant to list possible containment options and have the response team select from these options during the incident.

- Are there any technical controls that may not be in place? Is it possible that MFA was not enabled on all accounts? If so, these technical controls should be reviewed at this point.

Eradication

- What investigation can be done to concretely determine the scope of the compromise? We obviously need to understand all hooks the criminals have in the organization before we can be confident about an eradication plan.

- From the point of compromise, what other systems or data could the attacker touch? Are we confident we understand the extent of the breach?

- In addition to an IT investigation, it may be worth asking managers to discuss any other potentially suspicious or unusual behavior with their teams. While much of this may be noise, it can provide a comprehensive list of items to investigate before bringing everything back online.

Recovery

- Now that we believe everything is secure again, it is time to bring the compromised system, data, or accounts back online. Are there any last steps necessary to ensure the attacker does not regain access? This could be as simple as password changes, MFA resets, or implementing new technical controls.

- After everything is back up and running, what monitoring will be in place to quickly alert us if the attacker still has a foothold in the network?

Lessons Learned

- Everyone on the response team should discuss what went well and what didn't in a debrief. Incidents are challenging and there may be ways to improve the response plan after everyone has had experience with a real-world incident.

- What administrative and technical controls can be implemented to prevent similar incidents in the future? If they cannot be prevented, what can limit their scope?

Reporting

- Depending on the scope of the incident, communication should be made with stakeholders, team members, and possibly external contacts. We do not want to cause panic, but we want to assure them the incident is resolved.

- Legal and compliance resources should be consulted, as you many have legal or regulatory requirements for incident reporting.

- What internal documentation should be kept for future reference?

Once you have this plan written down and your management team is on board, you should be as prepared as possible for any potential incident. This puts you miles ahead of most organizations that have no plan.

CONCLUSION

WORK IN PROGRESS

We have come to the end of one road with this book, but your organization is just at the start of the cybersecurity journey.

Your business is a work in progress. It always has been. It always will be. The world is ever changing. Your cybersecurity program is no different.

Armed with a solid checklist of administrative and technical controls, you now know what the road map looks like. With full knowledge of the risks, you know what comes next. You also understand the questions to answer before an incident occurs.

If you'd like some assistance with your cybersecurity, my firm would be happy to help. You can take advantage of a special offer here:

https://cybersecuritychecklist.com/offer

We provide a full range of cybersecurity services, including security-first IT management, advanced security controls, cyber assurance services, and Virtual Chief Security Officer (vCSO) services. Depending on the level of risk and compliance needs, you may consider investing in these services in the future.

It is my sincere hope this book has shown you a path to more secure future. I wish you the best of luck and success in your business and in keeping cyber threats at bay.

To accelerate your progress and get exclusive resources that build on this checklist, please visit:

https://cybersecuritychecklist.com/extra

REFERENCES/SOURCES

Camarillo, Emmanuel. 2024. "800,000 People's Data Stolen in Lurie Children's Hospital Cyberattack." *Chicago Sun-Times*, July 2. https://chicago.suntimes.com/crime/2024/07/02/lurie-childrens-hospital-cybersecurity-breach-healthcare-tech-data-medical-records-files-fbi.

Censys Research Team. 2024. "Research Report: Internet-Connected Industrial Control Systems (Part One): Censys." Censys, August 7. https://censys.com/research-report-internet-connected-industrial-control-systems-part-one/.

Federal Communications Commission. 2022. "FCC Bans Equipment Authorizations for Chinese Telecommunications and Video Surveillance Equipment Deemed to Pose a Threat to National Security." Press release. November 25. https://docs.fcc.gov/public/attachments/DOC-389524A1.pdf.

Goodin, Dan. 2024. "GitHub Besieged by Millions of Malicious Repositories in Ongoing Attack." *Ars Technica*, February 28. https://arstechnica.com/security/2024/02/github-besieged-by-millions-of-malicious-repositories-in-ongoing-attack/.

Greenberg, Andy. 2024. "Russian Spies Jumped from One Network to Another via Wi-Fi in an Unprecedented Hack." *Wired*, November 22. https://www.wired.com/story/russia-gru-apt28-wifi-daisy-chain-breach/.

Hillman, Jonathan. 2022. "China Is Watching You." *Atlantic*, January 13. https://www.theatlantic.com/ideas/archive/2021/10/china-america-surveillance-hikvision/620404/.

Office of the Inspector General, Board of Governors of the Federal Reserve System. 2024. "Material Loss Review of Heartland Tri-State Bank." https://oig.federalreserve.gov/reports/board-material-loss-review-heartland-tri-state-bank-feb2024.htm.

Pistone, Ann, and Jason Knowles. 2024. "Woman Loses Nearly $1 Million Life Savings in 'Pig Butchering' Scam." *ABC30 Fresno*, September 9. https://abc30.com/post/woman-living-illinois-loses-1-million-life-savings-pig-butchering-scam-forced-sell-home-belongings/15271332/.

Rundle, James. 2024a. "MGM Seeks to Block FTC Probe of 2023 Cyberattack." *Wall Street Journal*, April 15. https://www.wsj.com/articles/mgm-seeks-to-block-ftc-probe-of-2023-cyberattack-2a2ca461.

Toulas, Bill. 2024. "Johnson Controls Says Ransomware Attack Cost 27 Million, Data Stolen." *BleepingComputer*, January 31. https://www.bleepingcomputer.com/news/security/johnson-controls-says-ransomware-attack-cost-27-million-data-stolen/.

U.S. Department of Justice. 2024. "Leader of Massive Scheme to Traffic in Fraudulent and Counterfeit Cisco Networking Equipment Sentenced to Prison." May 2. https://www.justice.gov/opa/pr/leader-massive-scheme-traffic-fraudulent-and-counterfeit-cisco-networking-equipment.

Washburn, Kaitlin. 2024. "Parents Sue Lurie Children's Hospital Over Cybersecurity Attack, Claim Hospital Failed to Keep Patients Safe." *Chicago Sun-Times*, July 19. https://chicago.suntimes.com/health/2024/07/18/lurie-childrens-hospital-chicago-cybersecurity-class-action-lawsuit.

Additional Works Consulted

Bošković, Marina M. Matić. 2023. "Cybercrime Money Laundering Cases and Digital Evidence." *Strani Pravni Zivot* 66 (4): 451–467. https://doi.org/10.56461/spz_22406kj.

Bressler, Martin S., and Linda Bressler. 2019. "Frauds, Embezzlers, Thieves, and Other Bad Actors: How Criminals Steal Your Profits and Put You Out Of Business." *Global Journal of Accounting and Finance* 3 (1). https://link.gale.com/apps/doc/A612476220/AONE?u=anon~c7e48f90&sid=googleScholar&xid=0e7474f1.

Cybersecurity Insiders, Inc. 2024. "2024 Cloud Security Report." https://engage.checkpoint.com/2024-cloud-security-report.

Cybersecurity Ventures. 2024. "Cybercrime to Cost the World $9.5 trillion USD annually in 2024." *Cybercrime Magazine*, November 18. https://cybersecurityventures.com/cybercrime-to-cost-the-world-9-trillion-annually-in-2024/.

Egress Software Technologies Inc. 2024. "Email Security Risk Report 2024." https://www.egress.com/media/o1sbpq5t/egress_email_security_risk_report_2024.pdf.

Electronic Privacy Information Center. n.d. "EPIC - Equifax Data Breach." https://archive.epic.org/privacy/data-breach/equifax/.

IBM. 2024a. "Cost of a Data Breach 2024." https://www.ibm.com/reports/data-breach.

———. 2024b. "IBM X-Force Threat Intelligence Index 2024." https://www.ibm.com/reports/threat-intelligence.

Mandiant. 2024. "M-Trends 2024 Special Report." *Google Cloud Security*. https://services.google.com/fh/files/misc/m-trends-2024.pdf.

Microsoft Corporation. 2024. "Microsoft Digital Defense Report 2024." https://cdn-dynmedia-1.microsoft.com/is/content/microsoftcorp/microsoft/final/en-us/microsoft-brand/documents/Microsoft%20Digital%20Defense%20Report%202024%20%281%29.pdf.

Palo Alto Networks. 2024. "Ransomware and Extortion Report 2023." https://www.paloaltonetworks.com/content/dam/pan/en_US/assets/pdf/reports/2023-unit42-ransomware-extortion-report.pdf.

Rundle, James. 2024b. "The AI Effect: Amazon Sees Nearly 1 Billion Cyber Threats a Day." *Wall Street Journal,* November 21. https://www.wsj.com/articles/the-ai-effect-amazon-sees-nearly-1-billion-cyber-threats-a-day-15434edd.

Woodruff-Sawyer & Co., Inc. 2024. "Looking Ahead: Cyber Insurance Trends for 2024." https://woodruffsawyer.com/insights/cyber-looking-ahead-guide.

William Clements has managed cybersecurity and information technology for small and mid-size businesses over the past 25 years. He has founded, co-founded, consulted, advised, partnered with, and invested in hundreds of businesses in dozens of industries. He founded the Chicago Technology Group, which is focused exclusively on keeping entrepreneurial businesses protected from cyber threats. He lives in Chicago, Illinois.

Take the next step in protecting your life's work:

https://cybersecuritychecklist.com/offer

www.ingramcontent.com/pod-product-compliance
Lightning Source LLC
LaVergne TN
LVHW020628100826
845148LV00012B/2100

* 9 7 8 0 9 7 6 2 2 4 8 1 5 *